Institute of Terrestria
NATURAL ENVIRONMENT

# A beginner's guide to
# Freshwater Algae

Hilary Belcher & Erica Swale
Culture Centre of Algae and Protozoa

London
Her Majesty's Stationery Office

© Crown copyright 1976
First published 1976
Third impression 1978

Culture Centre of Algae and Protozoa
36 Storey's Way
Cambridge
CB3 ODT

0223 (Cambridge) 61378

The Culture Centre of Algae and Protozoa
is part of the Institute of Terrestrial Ecology,
which is a component body of the
Natural Environment Research Council

ISBN 0 11 881393 5

# Introduction

The purpose of this booklet is to enable those who would like to learn to recognise some of the algae of freshwater to make a start. From the very large number of different genera we have selected 110 of those most likely to be found, and we have illustrated one species belonging to each. (A *genus*, plural *genera*, is comparable with the family name of a person eg 'Wordsworth', and a *species*, plural *species*, is comparable with a first name, eg 'William'.) The drawings are in almost every case made from living specimens (except those of diatom frustules).

WHAT ARE ALGAE?
Algae include such diverse groups of organisms that one may ask why they have been given the same name. The reason is that in the early days of the development of light microscopes, and before those of electron microscopes, their full diversity could not be detected. The cells of some algae are of bacterial size, for example, about 1 µm (1 micrometre = 1/1000 mm) or about 0.00004 ins across. Others contain a vast number of cells and can be up to 50 m (over 150 feet) long, for example, the largest seaweeds. Some algae are undoubtedly plants, some differ so little from bacteria that they could be members of a special bacterial class, some are animals or may show certain features both of animals and plants. Despite these facts most algal groups have certain features in common.

Microscopic algae can be found everywhere, from permanent snow and ice to deserts, the oceans, lakes, rivers, puddles, rocks and soil. Just as the trees, grasses and herbs everyone is familiar with are the main basis of life on land, so algae are the basis of that in the sea and, indeed, produce about the same amount of organic matter and oxygen as do land plants. Although there is nothing in freshwater comparable with the larger seaweeds, these algae show a range in size from the smallest of about 1 µm in diameter to trailing forms of more than a metre.

## HOW TO LOOK AT THEM

To examine freshwater algae the equipment required need not be elaborate. A few of the larger genera can be identified by a hand lens alone (magnifications of ×10 and ×20), but a microscope which magnifies at least ×100 is generally essential. At this stage there is no need at all for an oil immersion objective. It would take too much space, and it would not be very satisfactory, to explain in detail how to mount a sample of algae on a glass slide and how actually to use a microscope. For this a few minutes' demonstration by someone experienced is more valuable than pages of description.

Algae usually die and decay fairly rapidly after collection. To keep them fresh as long as possible, do not overfill the container, always leave an air-space and keep the sample cool. Plankton algae may be eaten by associated animals (such as *Daphnia*, the water flea) in a short time. If present in noticeable numbers they should be strained off immediately (a piece of nylon stocking makes a good sieve).

Samples may be preserved by adding iodine solution which, however, turns the chloroplast brown. A saturated solution of iodine in potassium iodide and water can be prepared, but ordinary tincture of iodine will do. It is best to add it from a dropping bottle until the whole is of a pale brown colour. It is important to ensure that the specimen bottle or tube of algae is kept well stoppered as iodine evaporates readily.

## NOTES ON THE DRAWINGS

Indication of sizes of the algae drawn is difficult as this varies so much between species, even of the same genus (for instance, cells of *Chlorogonium* range from 5–15 μm in length for *C. euchlorum* to 250 μm for *C. maximum*). It would be misleading to show the size of each species illustrated, so we have stated a range covering the whole genus.

The student is strongly advised to draw microscopic objects whenever possible. There is no necessity to 'be able to draw' to do this. What matters is having your own record of what the thing looked like, to help you to remember it and to understand its shape. It is surprising how trying to draw it makes one look much harder to see what the alga is really like. Always label the drawing with the name of the alga if known and add the date and the place where it was found.

CLASSIFICATION

To help find the way in more advanced publications (see page 46), the algae included here are arranged in their major groupings below:

Green algae (Chlorophyceae), including:
Swimming cells and colonies (Volvocales), (nos.1–14)
Non-motile cells and colonies (Chlorococcales), (nos.15–33)
Filamentous algae (of various families), nos.34–50)
Desmids (Conjugales in part), (nos.51–59)

Yellow-green algae (Xanthophyceae), (nos.60–63)

Golden-yellow algae (Chrysophyceae), (nos.64–68)

Diatoms (Bacillariophyceae), (nos.69–85)

Cryptophyceae, (no.86)

Dinoflagellates (Dinophyceae), (nos.87, 88)

Euglenoid flagellates (Euglenophyceae), (nos.89–91)

Red algae (Rhodophyceae), (nos.92–95)

Blue-green algae (Myxophyceae or Cyanophyceae), (nos.97–110)

# Glossary
(*see the diagram opposite*)

| | |
|---|---|
| *Axile* | in the centre of the cell |
| *Binary fission* | vegetative division of a free-living cell into two equal and separate daughter cells |
| *Chloroplast* | the coloured structure in plant cells which absorbs light used in photosynthesis; usually green or brown, occasionally red or blue-green |
| *Cyst* | resting-stage with thick wall, formed from a single cell |
| *Epiphytic* | used of a plant which grows on another, but not parasitically |
| *Flagella* | whip-like structures used by cells for swimming but difficult to see unless stained by iodine etc; hence 'flagellates'—cells swimming by flagella. (n.b. *flagellum* is singular, plural *flagella*) |
| *Frustule* | half-shell of a diatom—see note on diatoms following no.68 |
| *Girdle-view* | see note on diatoms and labelled diagram |
| *Nucleus* | colourless structure which contains hereditary apparatus of cell |
| *Parietal* | lying against the wall of a cell |
| *Plankton* | collective term for animals and plants which float in water and are carried passively by currents |
| *Pyrenoid* | a colourless structure associated with the chloroplast and concerned with starch formation |
| *Raphe* | the linear structure along the centre of some diatom frustules—see labelled diagram |
| *Sigmoid* | of an elongated S-shape, like *Gyrosigma* no.80 |
| *Stellate* | star-shaped, mainly used of chloroplasts |
| *Striae* | the stripe-like markings of a diatom frustule |
| *Valve* | see note on diatoms and labelled diagram |

# Diagrammatic structure of some algae

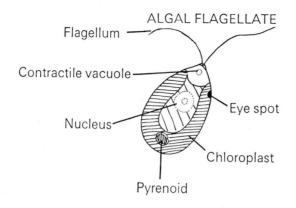
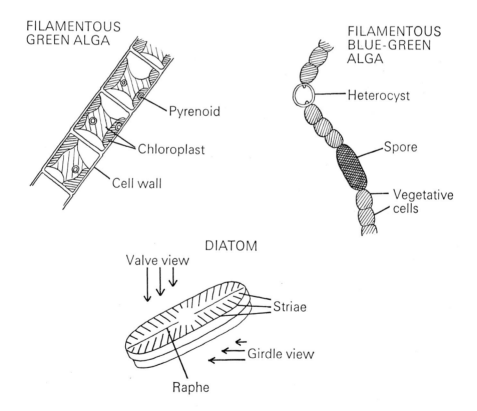

# Chlorophyceae : Green algae

Green swimming cells (see also *Euglena* and *Phacus*)

1 *Chlamydomonas*. A very big genus of round, oval or cylindrical cells which swim with two flagella of equal length. There may or may not be a pyrenoid, but a red eye-spot is usually present. The wall is smooth and thin. Cells can be from 2.5 to 50μm or more in length, mostly 5–20 (without including the flagella, as in all measurements in this booklet). Found in every kind of aquatic habitat, species of this genus are the most likely green swimming cells to be encountered.

2 *Chlorogonium*. This has two flagella like *Chlamydomonas* but is spindle-shaped. Unlike the similarly shaped *Euglena* (89) it does not change its form. 5–50μm long (1 rare species reaches 250μm), and found in ponds and rivers.

3 *Brachiomonas*. Related to *Chlamydomonas* but has a pointed tail and four lateral lobes which stick out of the body like fins. Sometimes found in bird-baths and in pools near the sea. 10–50μm long.

4 *Pteromonas*. Like *Chlamydomonas* but with a longitudinal wing on each side of the body like an elm fruit. Revolves as it swims, is 10–25μm long, and lives in ponds and slow rivers. 4b, a cell in transverse section.

5 *Haematococcus*. Cells oval, 10–30μm long, contents separated from the wall by a clear space crossed by radial threads. The green chloroplast is sometimes masked by red coloured oil. The dense populations of this organism occurring in bird-baths, wet hollows in rocks, the depressions in manhole covers and other small temporary water-bodies may be either bright green or rusty red.

6 *Lobomonas*. Cells like *Chlamydomonas*, 3–25μm long, with a lobed or wavy outline. Occurs in puddles, ponds etc, as does the similar *Diplostauron*, which is square.

7 *Phacotus*. Cells like *Chlamydomonas*, 10–20μm long, and found in similar places, but each is enclosed in a wall resembling two watch glasses fitting together, sometimes thin and transparent but more often thick and sculptured.

8 *Pyramimonas*. Cells oval or strawberry-shaped, 5–25μm long, with four flagella which arise from a pit at the front end. Anterior edge of the chloroplast deeply divided into four or eight lobes, giving the cell a striped appearance. There is no wall, and the cell divides by binary fission. Ponds, puddles and slow rivers.

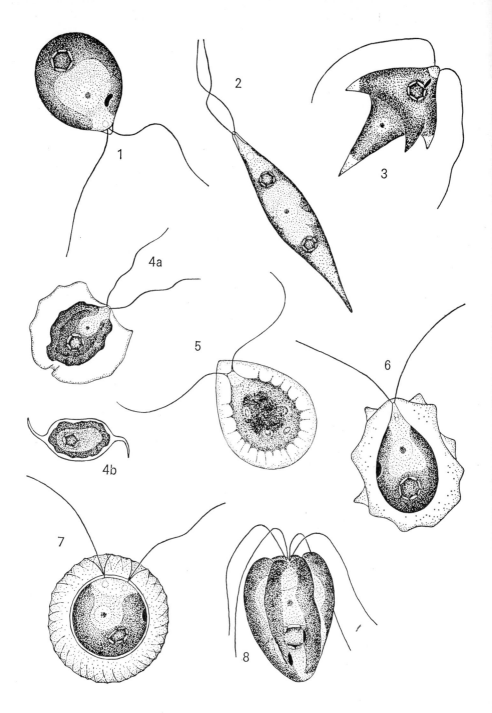

# Chlorophyceae : Green algae

Green swimming colonies

9. *Pyrobotrys*. Pear-shaped swimming cells 5–25 µm long, loosely aggregated to form a colony. In hoof prints and other puddles rich in organic matter. Will sometimes appear if mud and water from these is left to stand together with a piece of a dried pea. The similar *Pascherina* has only 4 cells instead of 8 or 16.
10. *Eudorina*. The small round cells, 5–15 µm long, occur at the surface of a globular colony up to 200 µm across. Several species, one common in puddles, ponds, lakes and rivers.
11. *Pandorina*. Differs from *Eudorina* in that the cells fit closely together without a large central space. Common in puddles, ponds, lakes and rivers. Cells 8–20 µm long, colonies up to 50 µm diameter.
12. *Gonium*. Consists of square, flat, four or 16-celled colonies up to 100 µm across. Cells 5–25 µm long. Common in puddles, ponds, lakes and slow rivers.
13. *Stephanosphaera*. The colonies are characteristic clear globes up to 60 µm across containing eight large cells like those of *Haematoccus* arranged in an equatorial ring. The globe revolves as it swims through the water. Found particularly in water-filled solution hollows of limestone rocks.
14. *Volvox*. Colonies visible to the unaided eye (up to 1 mm across), containing hundreds of cells. Found in puddles, ponds, lakes and slow rivers. For an article dealing with this and the previous four genera, see Jane, F. W., 1949. Famous plants: 1, *Volvox, New Biology*, 6, 87–99.

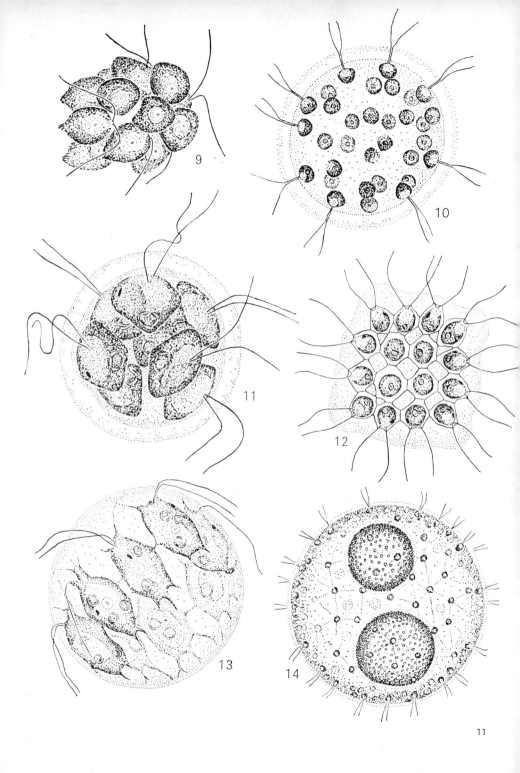

# Chlorophyceae: Green algae

Green non-motile unicells and colonies (but not Desmids)

15  *Chlorella*. Small round or oval cells (2–15 µm diameter) which divide into two or four non-motile daughter cells, enclosed for a little while within the old wall, as shown. Found everywhere, but sometimes occurring in vast quantities as a green soup in cattle-troughs, and similar places. For a fuller account see Fogg, G. E., 1953. Famous plants: 4, *Chlorella*. New Biology, 15, 99–116. *Chlorococcum* and its allies differ from *Chlorella* in producing motile spores, each with two flagella.
16  *Oocystis*. Oval or lemon-shaped cells 5–20 µm long, sometimes found 2 or 4 together inside the enlarged clear mother cell wall. Ponds, lakes and slow rivers.
17  *Chodatella*. Like *Oocystis* in size and shape, but with several slender bristles, usually at each end of the cell. Ponds, lakes and slow rivers.
18  *Tetraëdron*. Little angular green cells, 5–20 µm across, found in ponds, lakes and slow rivers. Two common species are illustrated.
19  *Ankistrodesmus*. Straight, curved or spiral needle-shaped cells up to 50 µm long or more, sometimes forming bundles. All types of water-body, common. The plate shows two species.
20  *Characium*. Oval or spindle-shaped green cells with a long or short stalk growing attached to various filamentous algae and other substrata. Length including stalk up to 80 µm. Common.
21  *Actinastrum*. Colonies of four or eight cigar-shaped cells, 10–25 µm long, united at one end to form a star. Ponds, lakes and slow rivers.
22  *Micractinium*. Round cells 3–20 µm across, each with one or more long slender bristles, often in colonies as shown. Plankton of lakes and slow rivers.

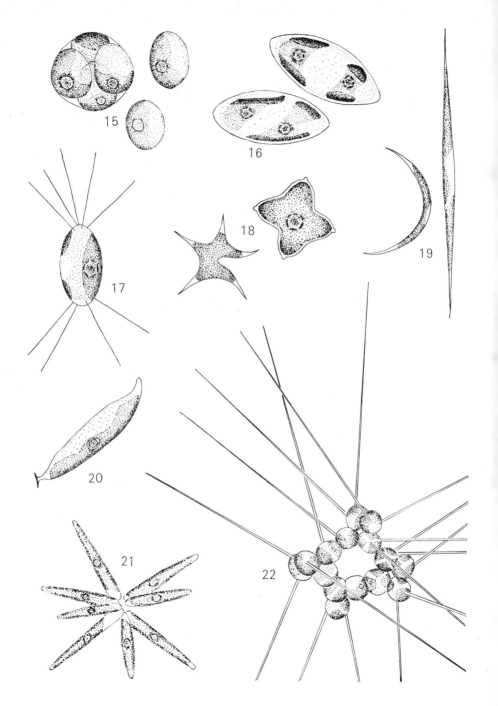

# Chlorophyceae: Green algae

Green non-motile unicells and colonies

23 *Coelastrum.* Non-motile roundish cells united closely to form spherical colonies up to 100 µm across. Plankton of lakes, ponds and slow rivers.

24 *Pediastrum.* Flat disc-shaped colonies, up to 100 µm across, of 4 to 32 cells, with a serrated edge and sometimes perforated. Ponds, lakes and slow rivers.

25 *Crucigenia.* Flat plates of oval cells united in groups of four, which are again often loosely united in fours (see also *Merismopedia*, no.99). Cells 3–15 µm long. Ponds, lakes and slow rivers.

26 *Scenedesmus.* Flat colonies of 2, 4 or 8 elongated cells arranged in a row as shown, some species having spines at the corners of the colonies. Cells 5–30 µm long. Common everywhere. Two species are illustrated.

27 *Tetrastrum.* 4-celled flat colonies arranged in a diamond as shown in the figure, often ornamented with small spines around the margin. Cells up to 10 µm across. Ponds, lakes and slow rivers.

28 *Dictyosphaerium.* Loose colonies of green cells in clear mucilage, the cells connected to a central point by branched strands (often barely visible). Cells up to 10 µm across. Ponds, lakes and slow rivers, sometimes very abundant.

29 *Hydrodictyon.* Net-like colonies of green cells, in rivers and canals. The cells, which become multinucleate, may grow up to 10 mm long, and a new net forms inside each. Well-grown nets are large and conspicuous. The figure shows a small portion of one.

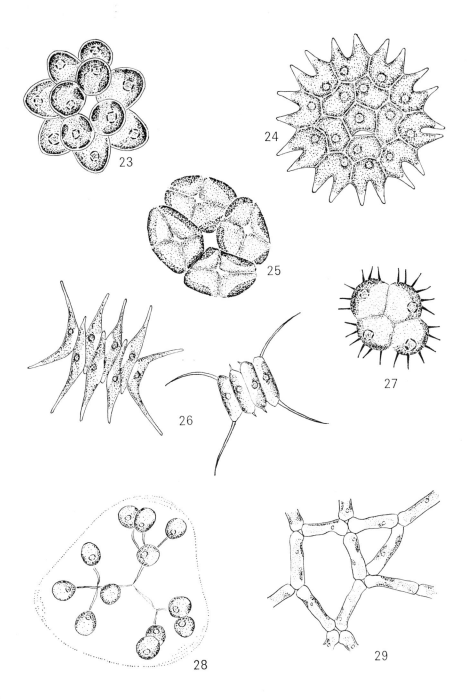

# Chlorophyceae

Green non-motile unicellular and filamentous algae

30  *Pleurococcus*. Green packets of cells very common on trees and fences, often in association with fungal hyphae. Also a constituent of some lichens. The cells are 10–20 µm across, and the 'hot cross bun' appearance is often characteristic.

31  *Trebouxia*. Spherical cells 2–25 µm diameter growing on trees, fences etc. and having a flat plate-like axile chloroplast with a lobed edge and a central pyrenoid. A constituent of some lichens. Sometimes difficult to distinguish from a species of *Chlorella* (15) that is also common on trees, but the *Chlorella* has a parietal chloroplast.

32  *Stichococcus*. A reduced filamentous form, with cylindrical cells 2–5 µm in diameter, growing singly or in little chains of 2 or 3 cells growing on trees, walls, gutters and flower pots, as well as in ponds or puddles. Very common.

33  *Botryococcus*. Grows in colonies up to 0.5 mm across, with numerous green cells 5–10 µm long embedded in an oily matrix which varies in colour from a clear yellow to totally opaque orange or brick red. Orange oil pigmented with carotene may be squeezed from it under a cover slip. Plankton of ponds and lakes, often floating on the water. Sometimes abundant, forming a thick scum. 33a, a colony; 33b, part of this flattened under a cover slip to show cells.

34  *Mougeotia*. Filamentous, unbranched, 5–50 µm diameter. Cells contain a single chloroplast, an axile ribbon running the length of the cell, with pyrenoids at intervals, seen in the wider view in the left-hand cell and in the narrower at the right. Common in standing water.

35  *Oedogonium*. Filaments unbranched, 3–50 µm diameter. Chloroplast a parietal net, with several pyrenoids. The thickened wall at one end of some cells, ridged like disposable plastic cups one inside the other, is characteristic (the 'cap'). Common in standing water.

36  *Zygnema*. Unbranched mucilaginous filaments 10–50 µm diameter, each cell with two star shaped chloroplasts with central pyrenoids. Common in standing waters. *Zygogonium*, found on damp peaty soil, is similar, but has purple sap.

37  *Spirogyra*. Unbranched mucilaginous filaments, varying from 10 to over 100 µm diameter, cell with one or more parietal chloroplasts in the form of spirally wound ribbons. Common in standing water, especially in early spring.

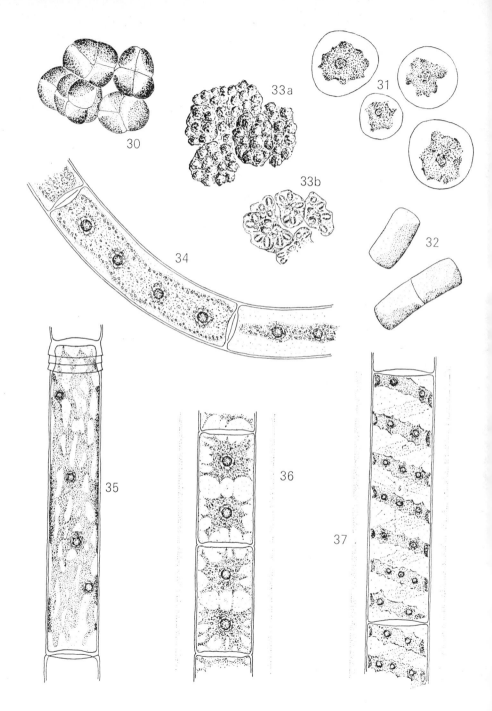

# Chlorophyceae
## Green filamentous algae

38 *Ulothrix*. Unbranched filaments 5–70 µm diameter, each cell with one chloroplast (or 2 in dividing cells) which forms an incomplete band round the cell as shown. Common in standing or running water, especially in spring. If branched, see *Stigeoclonium* (44) or *Draparnaldia* (40).

39 *Microspora*. Unbranched filaments 5–25 µm diameter. Cells appear to have numerous parietal chloroplasts (though this is not the case), and are somewhat similar in structure to *Tribonema* (61), with H-pieces (overlapping half-cell walls, illustrated for *Tribonema*), but are relatively shorter and thicker, and are a more opaque darker green. In running or standing water. The upper two cells in the figure are seen in optical section, the lower two in surface view.

40 *Draparnaldia*. Cells resemble those of *Ulothrix* with a band-shaped chloroplast, but the whole plant has a feathery look, with a central axis of large cells, 50 µm or so across in the older parts, bearing branched tufts of much narrower cells at intervals. Standing and running water, especially in the spring, forming conspicuous bright green tufts up to 5 cm long on stones etc.

41 *Bulbochaete*. Small plants with branched filaments 10–30 µm diameter. Cells have a net-like chloroplast resembling that of *Oedogonium* (35), and many bear long hairs with bulbous bases. They grow on other algae, stones, woodwork, etc, standing out stiffly. Usually in slowly flowing or still water. 41a, two cells; 41b, habit sketch.

42 *Cladophora*. Filaments up to 100 µm across, usually branched cells with reticulate chloroplasts. Plants often large and rough to the touch, frequently covered with epiphytes. Common, usually attached to sticks, stones or other hard surfaces in still or running water. Is very troublesome to pond owners, as it rapidly grows into skeins and tangles several metres long, and it is known to them as 'blanket weed'. 42b, habit sketch. Certain species form furry balls up to the size of tennis balls, which roll around on the bottom of shallow lakes and pools, and are recorded from various parts of Britain.

Filamentous algae are often found which have the cell structure of *Cladophora* but are unbranched. These may belong either to *Cladophora* or *Rhizoclonium*, and deciding which is often difficult.

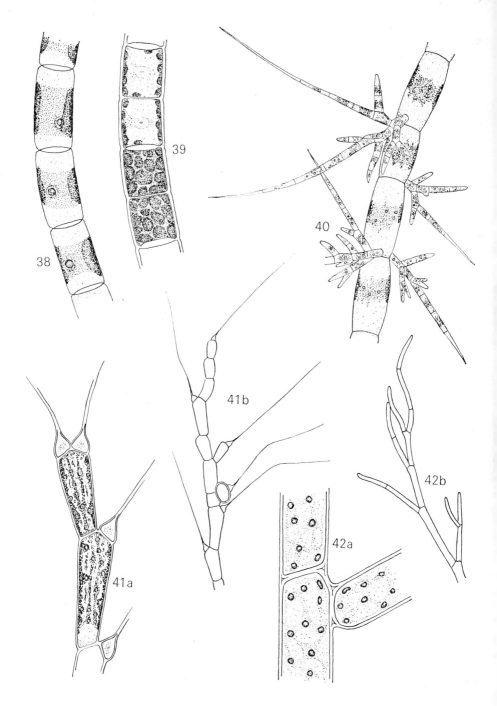

# Chlorophyceae

Filamentous, sheet-like or bag-like green algae

43 *Prasiola*. Cells 7–20 µm across, having an axile chloroplast in the form of a plate with a central pyrenoid, like *Trebouxia*. The plant may be filamentous (43a), a ribbon several cells wide (43b, smaller scale), or expanded into a sheet (43c, natural size). Aquatic or subaerial, growing on rocks frequented by birds, or in other habitats rich in nitrogen.

44 *Stigeoclonium*. Cells like those of *Ulothrix*, with band-shaped parietal chloroplasts. Filaments branched, up to 20 µm across or more, but without a stout central axis as has the related *Draparnaldia* (no.40), and considerably smaller and softer than *Cladophora* (no.42). Common in standing or running water, especially in spring. 44a, two cells about to divide (with two chloroplasts each) ; 44b, habit sketch.

45 *Chaetophora*. Branched filaments growing into woolly tufts or balls (45b, ×10), from which project long colourless hairs. Cells up to 20 µm across. Common as an epiphyte of water plants, or on sticks or stones, sometimes encrusted with crystals of calcium carbonate.

46 *Enteromorpha*. Plant body tubular, hollow (46b, c, natural size), up to 1 cm wide and 15 cm or more long, composed of cells with thick walls and parietal chloroplasts (46a). Sometimes very conspicuous floating in masses on the surface of slow rivers and canals, or entangled with larger water plants, and may be abundant in brackish water.

47 *Tetraspora*. Put here because of the macroscopic plant bodies it forms (47b, slightly enlarged), but these are of mucilage containing cells of *Chlamydomonas* type, having pseudocilia (like motionless flagella) projecting from the mucilage (47a). *Palmodictyon* is similar, but forms reticulate jelly-masses.

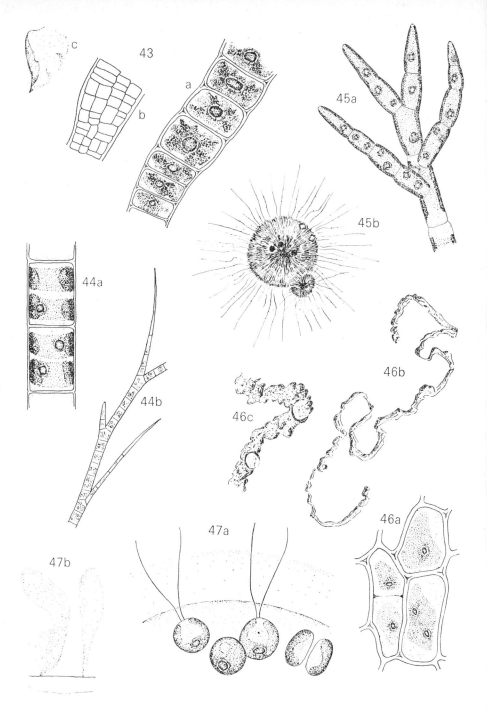

# Chlorophyceae

Filamentous and sheet-like green algae and Desmids

48 *Aphanochaete*. Small filamentous forms 5–15 μm in diameter, which grow prostrate upon other algae. The cells have parietal chloroplasts, and some of them bear hairs with bulbous bases. 48a, two cells, one with hair; 48b, habit sketch, growing on *Oedogonium*.

49 *Trentepohlia*. Although belonging to the green algae this branched filamentous plant has cells filled with orange oil. It is subaerial and grows on damp rocks, forming patches which look like orange velvet. Much commoner in northern and western Britain than in the south-east. Cells 7–20 μm diameter.

50 *Coleochaete*. This forms disc-shaped encrusting plates of cells beset with fine hairs, and grows on stones and waterplants, reaching a diameter of 1 mm, while the cells are 10–20 μm across. It is often conspicuous on the glass of aquaria or the sides of old sinks in which water plants are grown. 50a, details of cells; 50b, part of a large plant at a lower magnification; 50c, a whole young plant.

DESMIDS (A group of specialised green algae)

51 *Hyalotheca*. A representative of the small number of filamentous desmids which are found in boggy pools etc. The filaments are cylindrical, 5–40 μm in diameter, and surrounded by a layer of mucilage.
*Desmidium* is a similar genus, but the filaments usually have a toothed outline and are most frequently triangular in cross section.

52 *Mesotaenium*. A small cylindrical desmid, 8–30 μm diameter, with a flat chloroplast like *Mougeotia*. It is particularly common in the jelly-like covering of wet rocks down which water trickles, and is usually accompanied by blue-green algae of various species. *Cylindrocystis* is another small desmid which is very similar but this has two separate stellate chloroplasts.

53 *Pleurotaenium*. A rod-shaped desmid up to 1 mm in length, found in boggy places and upon wet rocks down which water runs. The wavy outline near the centre of the cell is characteristic.

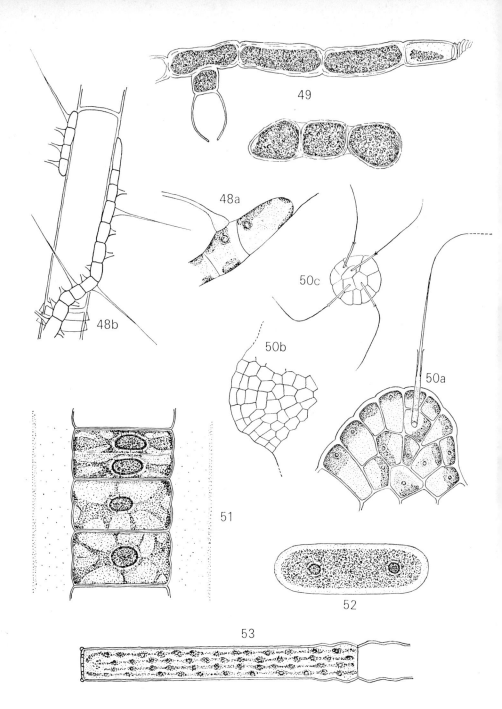

# Chlorophyceae : Green algae

Desmids

54 *Closterium*. A crescent-shaped desmid found commonly in ponds, lakes and slow rivers, as well as in bogs. Some of the species have large and conspicuous cells reaching nearly 1 mm in length, while others, particularly the planktonic ones, may be much more slender and almost straight.

55 *Cosmarium*. Elliptical or angular in outline, but without spines or other processes. A big and difficult genus, particularly common in boggy places, but a few species are found in water bodies of all types. 10–200µ long.

56 *Staurastrum*. Each half cell is triangular in end view, with three corners, spines or arms. Cells vary up to 130 µm in length, and are common in the plankton of lakes.

57 *Micrasterias*. Cells very flat and leaf- or plate-like, oval in outline, with the edges variously cut of fissured. Up to 350 µm long. Acidic pools and boggy places.

58 *Xanthidium*. Cells ornamented with various combinations of spines, but not triangular in end view as *Staurastrum*. 10–200 µm long. Found in acidic or boggy pools etc.

59 *Euastrum*. Cells somewhat flattened, but not to such a degree as in *Micrasterias*, and having angular lobes, while the ends are often deeply notched. Cells 10–200 µm long. Acidic pools and boggy places.

*A note on desmids*

Only a few tolerant members of this group are found in the hard calcareous waters of the south and east of Britain. The vast majority are confined to districts of acidic rocks and soft water, and while some of these live in the plankton of lakes, most are found in boggy pools, where there is often an abundance of species and individuals. A rather old but still useful account, with many coloured plates is West, W. & West, G. S., A Monograph of the British Desmidiaceae. 5 volumes. Ray Society, London, 1904–1923.

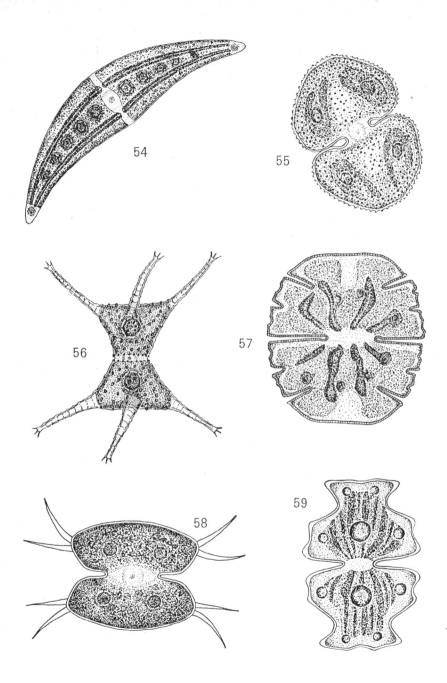

## Xanthophyceae

### Yellow-green algae

60 *Ophiocytium.* Cylindrical or sausage-shaped epiphytic or free-floating unicells, 3–20 μm diameter, each with two or more pale green chloroplasts. They vary from short to very elongated, and are often curved or even loosely coiled. In the common species shown here the young cells attach themselves to the lip of the mother cell wall, which is itself attached to a *Cladophora* filament.

61 *Tribonema.* Filamentous and unbranched, 5–20 μm diameter, with several parietal chloroplasts, and H-pieces, resembling *Microspora* (39), but differing in the longer and narrower cells and much paler and more transparent appearance, having no starch. Common, especially in grassy pools. 61a part of a filament; 61b, H-piece (fit together to form the walls).

62 *Botrydium.* A coenocytic (multinucleate) alga forming dark green globules up to 2 mm across on drying mud, anchored by a colourless branched rhizoid. Not uncommon, but only lasts in this form for a few days, perennation being by cysts. 62a, a whole plant dug up, ×10; 62b, group of plants on mud, seen from above, ×2.5.

63 *Vaucheria.* A very common bright green filamentous alga, especially in stagnant or slowly running water. Plant coenocytic, of even thickness throughout (50–200 μm), irregularly branched, and with no cross walls. Numerous small green chloroplasts. 63a, end of a filament; 63b, antheridium (male cell) and two oogonia (female cells).

## Chrysophyceae

### Golden-brown algae

64 *Ochromonas.* Motile unicell, usually oval, 2–25 μm long, with 1 or 2 brown chloroplasts and a long and a short flagellum. *Chromulina* is similar but only has 1 flagellum visible. Like many of their class they do not live long on a microscope slide. Fairly common, especially in spring, in puddles, ponds, lakes etc.

65 *Mallomonas.* Like *Ochromonas,* but covered all over with glassy scales of silica, overlapping like roof tiles. Some species also have long silica spines, at one or both ends or all over (as illustrated). Cells 5–50 μm long. Fairly common in puddles, ditches, ponds and lakes, especially in spring.

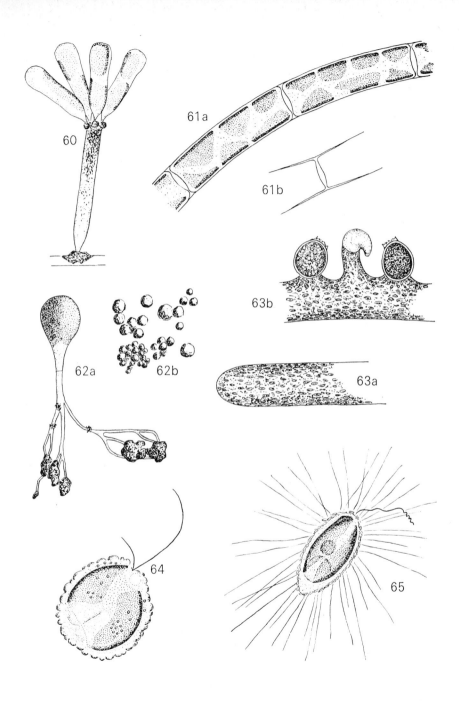

## Chrysophyceae

continued

66   *Dinobryon*. Brown cells like *Ochromonas*, but each living in a clear lorica (pot). Solitary or forming branched colonies, attached or free-swimming. Length of lorica 30–100 µm. Frequent in ponds and lakes, especially in spring.
67   *Synura*. Scaly brown cells like *Mallomonas*, 20–40 µm long, connected by their tails into swimming colonies up to 500 µm across. Common in puddles, ponds, lakes and slow rivers. 67a, swimming colony; 67b, single cell, larger.
68   *Uroglena*. Forms swimming colonies up to 500 µm across, but the small cells sit at the edge of a ball of soft clear almost invisible mucilage. The tails go right to the centre, but cannot often be seen. Common in lakes, pools etc. in spring. 68a, a colony; 68b, 3 cells from its edge; 68c, silica-walled cyst or resting stage. Many members of the Chrysophyceae produce these round or oval bottle-like cysts.

## Diatoms or Bacillariophyceae

Brown algae with silica shells

(The shell or frustule is like a box, formed of 2 overlapping glassy portions as in a Petri dish or a date box. The side view of the cell is the girdle view and the top or bottom view the valve view (see diagram, p. 7). In the latter the characteristic sculpturing of the wall can be seen, especially in frustules cleaned with acid or heat and specially mounted as is needed for detailed identification. Methods in Lund, 1961.)

69   *Melosira*. Cylindrical cells joined together to produce filaments, 5–100 µm diameter, with numerous brown chloroplasts. Common in ditches, ponds lakes. The species illustrated (in girdle view) has smooth walls, but others bear rows of spots, and sometimes also spines.
70   *Cyclotella* and *Stephanodiscus*. Often these cannot easily be told apart, but only by detailed study. They are cylindrical diatoms, 5–50 µm diameter, occurring singly or in pairs but not in long chains. Often they have a crown of very fine long radiating bristles from each end, not shown here because it is difficult to see. Common, usually planktonic, in pools, lakes, and especially slow rivers, whose waters are often brown with them in spring and early summer. 70a, b, valve and girdle views; 70c, cleaned valve of a *Cyclotella*.

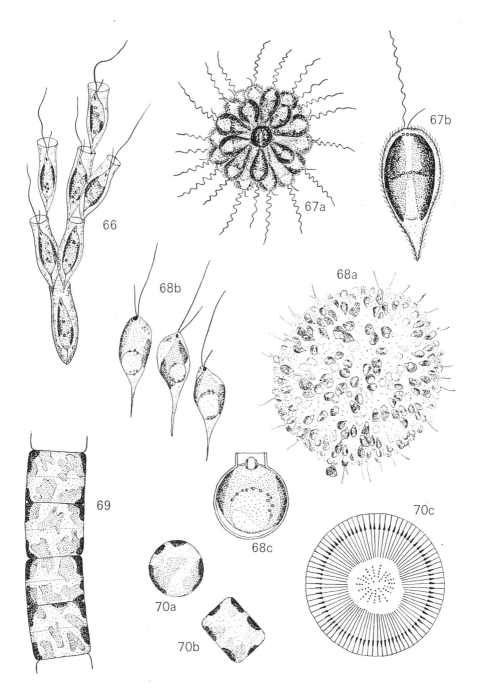

# Diatoms

continued

71 *Tabellaria*. A diatom forming flat zig-zag filaments as illustrated, or else radial colonies like those of *Asterionella* (76). The characteristic septa extend part way across the cells in girdle view and show up as heavy lines. Length of cells, up to 100 µm or more. 71a, part of a zig-zag chain of cells, held together by mucilage pads, in girdle view. The brown chloroplasts have only been drawn in the upper cell. 71b, valve view of one of these cells. More common in the less calcareous waters of the north and west.

72 *Diatoma*. Another diatom which forms filaments, straight or zig-zag. Cells up to 70 µm long, without septa in girdle view, unlike the last genus. In valve view (72a) they vary from linear to broadly lanceolate, but are recognised by their irregular transverse ribs. Common in water bodies of all types. 72a, valve view, cleaned frustule; 72b, girdle view of living filament.

73 *Meridion*. This has cells like *Diatoma* (up to 100 µm long), but wedge-shaped, and forms characteristic fan-shaped colonies. Common, especially in spring, in ponds, brooks and rivers. 73a, two cleaned frustules, girdle view; 73b, another in valve view; 73c, girdle view of living colony.

74 *Synedra*. A long narrow diatom, not forming chains, occurring in the free state (often planktonic) or attached to a substratum by a pad of mucilage to form radiating colonies. Cells up to 500 µm long. Common. 74a, valve view of cleaned frustule; 74b, living colony, girdle view.

75 *Fragilaria*. Resembles *Synedra* in valve view, but cells (up to 150 µm long) form long flat filamentous colonies, benthic or planktonic. 75a and 75b show filaments of two species of very different appearance. Common in ponds and lakes.

76 *Asterionella*. Forms star-shaped floating colonies of cells up to 100 µm long, and is common in lakes. There is also a form of *Tabellaria* (71) which has colonies similar in size and shape but whose cells are recognised by their longitudinal septa and by the knee-like swellings at the centre of each cell.

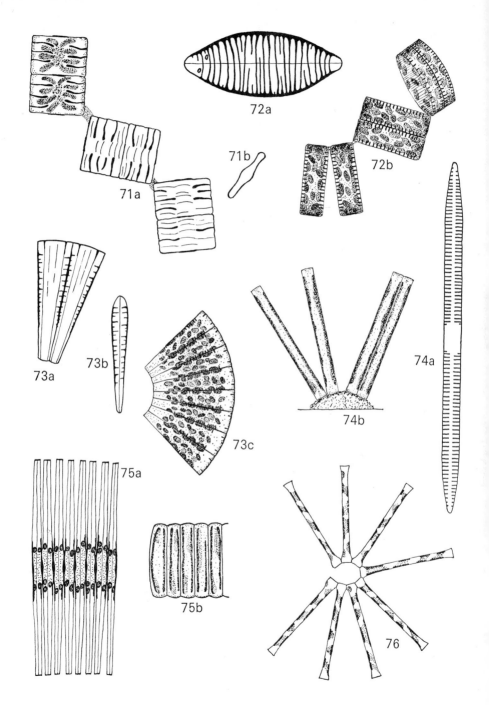

# Diatoms

continued

77 *Cocconeis*. Extremely common flat oval diatoms, 10–30 µm long, clinging to plants, stones and other substrata, sometimes in an almost continuous layer. 77a, cleaned frustule; 77b, 3 cells on an *Oedogonium* filament.

78 *Navicula*. Elongated diatoms, with many species varying from 5 to 200 µm in length, creeping on mud etc in all types of water-body. Various related genera (*Achnanthes* etc) only distinguishable from *Navicula* by detailed examination. 78a, cleaned frustule; 78b, living cell with a chloroplast on each side. Girdle view parallel-sided.

79 *Pinnularia*. Resembles a large *Navicula* with thick bar-like striae and rounded ends, 20–400 µm long. On the mud of ponds, lakes and rivers. The figure shows a cleaned frustule.

80 *Gyrosigma*. Recognised by its beautiful sigmoid shape, is up to 200 µm long, and lives on the mud or lakes, ponds and rivers. See also no. 83. The figure shows a cleaned frustule.

81 *Cymbella*. Shaped like a Cornish pasty, 10–200 µm long, commonly attached to plants and other substrata by mucilage stalks, or living several together in mucilage tubes. 81a, cleaned frustule; 81b, living cell on stalk of mucilage. *Amphora* looks like two *Cymbellas* joined to form an oval.

82 *Gomphonema*. Also common, usually attached by mucilage stalks, but can also creep about like the other diatoms on this page and the next. Shape asymmetrical lengthwise, girdle view wedge-shaped, and valve view varying from narrowly oval (not elliptical) to the shape of an Egyptian mummy-case, 8–100 µm long. 82a, cleaned frustule; 82b, c, living attached cells, valve and girdle views. A diatom similar to *Gomphonema* but curved in girdle view like a banana is *Rhoicosphenia*.

83 *Nitzschia*. Common, many species, 5 to more than 100 µm long. Best distinguished in live state from *Navicula* and its allies by the two chloroplasts at the ends of the cell, not the sides. There is also a big sigmoid species, up to 600 µm long, but unlike *Gyrosigma* (80) it has parallel sides and blunt ends. 83a, cleaned frustule, 83b, living cell, both of a small species.

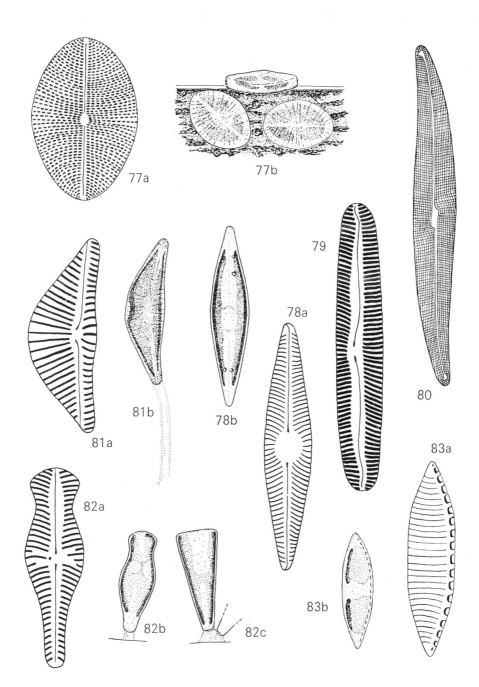

84 *Surirella*. Large benthic diatoms, valve view elliptical or oval, girdle view rectangular or wedge-shaped, 15–200 µm long. Edge of valve often forms balustrade-like 'wing'. Ponds, lakes and rivers. 84a, valve view ; 84b, girdle view.

85 *Cymatopleura*. Resembling *Surirella* and found in similar places, but cells much flatter, like biscuits, and lacking the ornamental 'wing'. Two common species, one constricted in girdle view as shown, the other elliptical. 30–150 µm long.

## Cryptophyceae

A small group of flagellates, mostly brown to olive green

86 *Cryptomonas*. Asymmetrical heart-shaped or oval cells, 10–80 µm long, with brownish chloroplasts, two unequal flagella and a gullet lined with structures called 'trichocysts', showing up as rows of spots. Common in water-bodies of all types.

## Dinophyceae

A small group of flagellates, usually armoured

87 *Ceratium*. Brown, armoured and drawn out into characteristic horns as shown. Two flagella, one longitudinal, the other transverse and in a groove. Up to 400 µm long. Sometimes common in lakes and pools.

88 *Peridinium*. Armoured, helmet-like, brown or colourless, with two flagella, 15–70 µm across. Fairly common in lakes and pools, recognisable by lumpy shape.

## Euglenophyceae

A group of green flagellates

89 *Euglena*. Spindle-shaped body, can become fatter or thinner, with one flagellum, green chloroplasts and red eye spot. 25–100 µm long. In all water bodies, especially slightly polluted ones, sometimes colouring farm ponds dark green.

90 *Phacus*. Also has green chloroplasts, bright red eye spot, and single flagellum, but is flattened like a leaf, with a pointed tail. Length up to 100 µm. In the same habitats as *Euglena*.

91 *Trachelomonas*. The green chloroplasts are usually masked by the brown round or oval pot, up to 25 µm across in which it lives, the flagellum protruding through a hole. Common in all water bodies. 91a, a cell in its pot ; 91b, more highly magnified, pot opened to show flagellate inside.

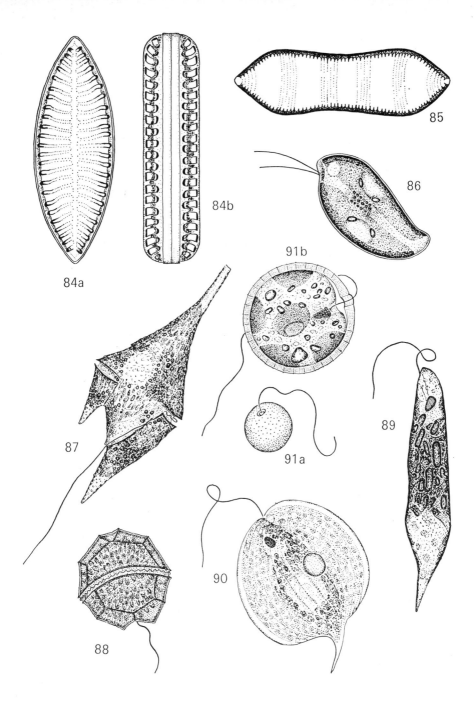

## Rhodophyceae

'Red' algae (often brown or blue-green in colour, however, especially in fresh water)

92 *Porphyridium*. Bright red unicells growing at the foot of walls and rocks and in greenhouses, where soluble salts collect, forming mucilaginous masses resembling drops of blood or tomato sauce. Occasional. Cells are about 10 μm in diameter and have a stellate chloroplast.

93 *Asterocytis* (or *Chroodactylon*). Forms short filaments or cushions of cells about 10 μm across, each with a bright blue-green stellate chloroplast and a central pyrenoid. Often epiphytic on other algae, especially *Cladophora*, and fairly common in hard water districts.

94 *Batrachospermum*. Highly organised filaments, forming brown or green mucilaginous branching tufts up to several cm long on rocks and stones in streams and springs, occasionally in mountain tarns. Branches usually consist of a central axis from which arise whorls of tiny branches, usually giving a beaded appearance. 94a, axis with whorls of small branches and two major ones (the 'birds' nests' are reproductive bodies); 94b and c, small branches, at a moderate and a high magnification.

95 *Lemanea*. Another highly organised form up to 10–15 cm long. Plants are thick strands, knobbly and sparsely branched, dull olive green, brownish or grey. Commonly forms large tufts on stones in shallow fast rivers, especially in spring.

## Myxophyceae (or Cyanophyceae)

Blue-green algae (which have no readily distinguishable nuclei or chloroplasts). Related to bacteria.

96 *Chroococcus*. Spherical or elliptical cells occurring singly or in pairs or fours, occasionally more, dividing in 3 planes at right angles to each other and surrounded by stratified sheaths. Contents granular, blue-green or brownish. Not uncommon in bog pools and salt marshes. Cells up to 50 μm across, but usually half or a quarter this size.

97 *Gloeocapsa*. Similar to *Chroococcus*, but cells in general much smaller, up to 8 μm across, in groups in wide stratified mucilage sheaths, as shown, sometimes bright red, forming masses on wet rocks.

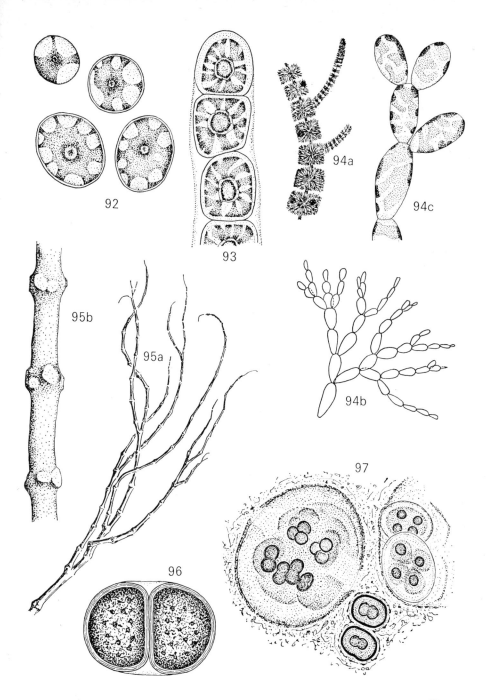

# Myxophyceae

Blue-green algae

98 *Aphanothece*. Small elongated cells, 3–8 μm long, embedded loosely in transparent structureless mucilage, mostly on wet rocks and in boggy pools. A similar genus but with round cells is *Aphanocapsa*. *Synechococcus* is another genus which has single cells like those of *Aphanothece* (not in colonies). Part of a colony of *Aphanothece* is shown.

99 *Merismopedia*. Easily recognised by the round cells, 3–10 μm diameter, dividing only in two planes, producing flat colonies like rafts, a small one being illustrated. Not uncommon in ponds, lakes, bogs, slow rivers.

100 *Gomphosphaeria* (*Coelosphaerium*). Small cells up to 5 μm in diameter, united into a hollow, roughly spherical, floating colony, sometimes blackish in appearance and often surrounded by mucilage. Colonies reach about 100 μm across. Conspicuous in the plankton of lakes.

101 *Microcystis*. Small cells up to 5 μm in diameter, embedded densely in mucilage to form irregular colonies, often pierced with holes. In bogs and lakes, sometimes planktonic. *Lamprocystis*, one of the purple sulphur bacteria, has colonies like *Microcystis*, but is pink, and lives in ponds and ditches with much decaying vegetation.

102 *Chamaesiphon*. Small elongated cells up to 5 μm diameter attached by one end to larger filamentous algae etc, while from the other end a chain of spores is produced. Common. The figure shows three plants of different ages on a filament of *Cladophora*.

103 *Oscillatoria*, *Lyngbya*, *Phormidium*. Three genera difficult to tell apart, even by specialists. Cells united into unbranched filaments, without spores or other specialised bodies, but often with a differently shaped end cell. The filaments can creep slowly along. Diameters vary from 1 μm to 30 or more. Sometimes found as single filaments, sometimes forming dense mats, often dark bluish green, olive or purplish brown in colour, and with a characteristic smell. Very common in all sorts of water bodies, either planktonic or benthic, also in air on wet rocks and soil.

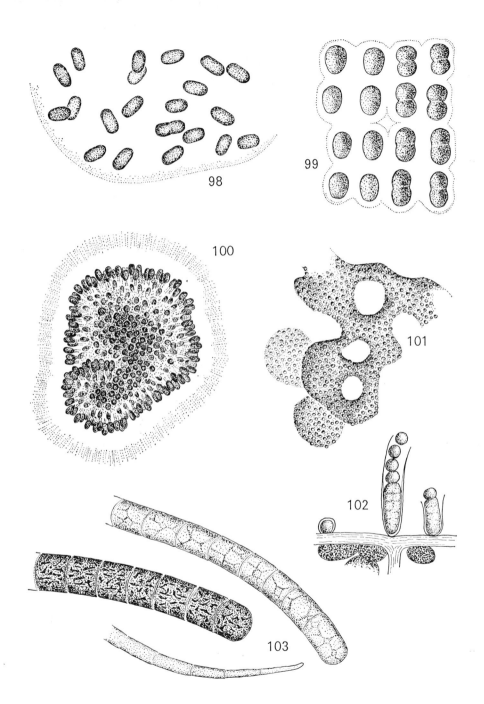

# Myxophyceae

continued

104 *Anabaena*. Planktonic filamentous algae recognisable by their beaded appearance, often curly, and sometimes having thick-walled clearer 'heterocysts' (specialised cells) and sausage-shaped dense spores at intervals. Cells up to 10 µm across. Common in the plankton of lakes and ponds, but sometimes occurring in bogs. The figures show parts of two filaments at different magnifications.

105 *Nostoc*. Filaments like those of *Anabaena*, but aggregated closely together to form a colony of definite shape which may be green or blackish. On the beds of rivers, on wet rocks, wet soil and flower pots. Cells up to 9 µm across, colonies up to several centimetres. 105a, filaments inside a young colony; 105b, a colony, natural size.

106 *Aphanizomenon*. Filaments somewhat like those of *Anabaena*, but less beaded, 2–6 µm diameter, aggregated together into bundles up to several mm long which float at the surface of ponds and lakes, sometimes abundantly. 106a, parts of 3 filaments, showing a spore (above) and a heterocyst (below); 106b, a floating colony, ×20.

107 *Gloeotrichia*. Cells in tapering filaments with a heterocyst at the thick end. Filaments clustered together by the blunt ends forming spherical furry balls up to several mm diameter, floating at the surface of ponds and lakes. 107a, several filaments; 107b, part of these, larger, the right-hand one with a spore; 107c, a floating colony, ×20.

*A note on water-blooms*

Some of the planktonic blue-green algae at times produce very large floating masses at the surface of pools and lakes in calm weather, particularly in autumn. These are very conspicuous, often appearing like spilt green paint, and have been known for many years as water-blooms. (See the article by A. J. Brook, 'Water-Blooms', in *New Biology*, vol. 23, 1957.) Genera liable to form blooms include *Gomphosphaeria*, *Microcystis*, *Oscillatoria*, *Anabaena*, *Aphanizomenon*, *Gloeotrichia*. The green colonial alga *Botryococcus* also forms blooms at times.

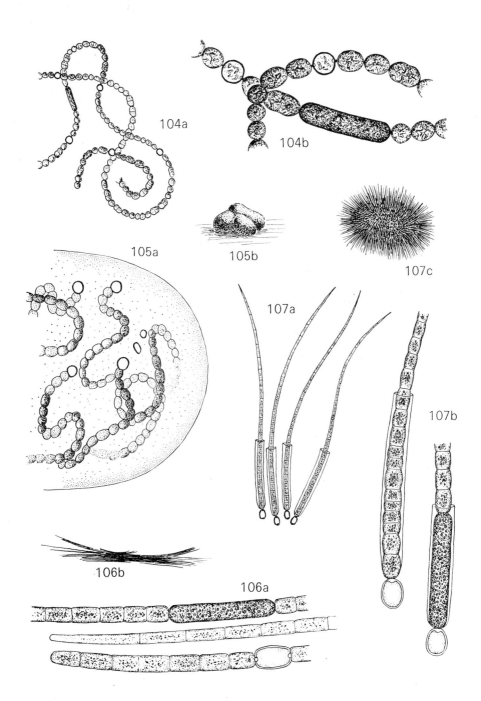

# Myxophyceae

## Blue-green algae

108 *Tolypothrix*. This filamentous form has branches of a characteristic type, as shown, usually marked by heterocysts. It occurs among other water plants. The cells are about 10 μm broad. 108a, filament with characteristic branch; 108b, end of a filament.
*Scytonema* is similar to *Tolypothrix*, but the sheaths are much wider and it grows on wet rocks or soil surfaces.

109 *Rivularia*. The tapering filaments are up to 12 μm in diameter, and are much branched, with heterocysts at the bases of the branches. These filaments are united to form tough hemispherical colonies several mm in diameter, especially conspicuous on stones in mountain streams, but also occurring on reeds and other stems in pools. 109a, colonies on a stone, natural size; 109b, part of a branched filament; 109c, filaments more highly magnified.
*Calothrix* has filaments like *Rivularia*, but is much less branched, and the filaments occur singly or in tufts rather than colonies.

110 *Stigonema*. The filaments branch irregularly, and usually form dark furry masses on damp rocks. The cells are rounded, and lie several abreast in the older parts, as shown. Heterocysts scattered, but are not regularly at the bases of branches. The cells vary from 3 to 30 μm in diameter, but of course the older parts of the plant are very much wider than this. 110a, part of whole plant, ×20; 110b, tip of a plant showing irregular rows of cells surrounded by mucilage.

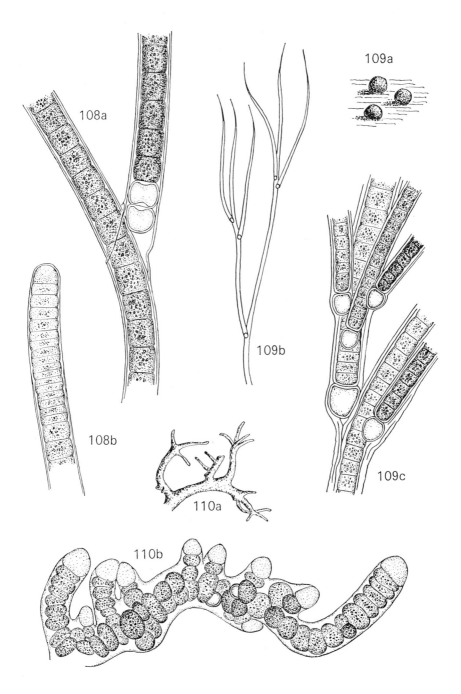

# Notes

# Notes

# Further reading

After becoming familiar with the algae in this booklet you may want to learn more. There is a large specialist literature dealing with algal identification. Fortunately, unlike many groups of plants and animals, freshwater algae are mostly world-wide in their distribution, so that, for example, American and German floras will be found useful.

For the almost-beginner:

Prescott, G. W. *How to know the Freshwater Algae.* Brown, Iowa, 1954

West, G. S. & Fritsch, F. E. *British Freshwater Algae,* Cambridge, 1927

For a general introduction:

Round, F. E. *The Biology of the Algae.* Edward Arnold, London, 1973

The following two articles give help on collection and examination:

Lund, J. W. G., 1961. The algae of the Malham Tarn district. *Field Studies,* 1(3), 85–119. (Here the hints on collection appear after the references.)

Lund, J. W. G., 1960. The microscopical examination of freshwater. *Proceedings of the Society for Water Treatment and Examination,* 9(2), 109–144.

Indispensable for the serious student:

George, E. A., 1976. A guide to algal keys (excluding seaweeds). *British Phycological Journal,* 11(1), 49–55.

# Index

Actinastrum 21
Achnanthes 78
Amphora 81
Anabaena 104
Ankistrodesmus 19
Aphanizomenon 106
Aphanocapsa 98
Aphanochaete 48
Aphanothece 98
Asterionella 76
Asterocytis 93

Batrachospermum 94
Botrydium 62
Botryococcus 33
Brachiomonas 3
Bulbochaete 41

Calothrix 109
Ceratium 87
Chaetophora 45
Chamaesiphon 102
Characium 20
Chodatella 17
Chlamydomonas 1
Chlorella 15
Chlorococcum 15
Chlorogonium 2
Chromulina 64
Chroococcus 96
Chroodactylon 93
Cladophora 42
Closterium 54
Cocconeis 77
Coelastrum 23
Coelosphaerium 100
Coleochaete 50
Cosmarium 55
Crucigenia 25
Cryptomonas 86
Cyclotella 70
Cylindrocystis 52
Cymatopleura 85
Cymbella 81

Desmidium 51
Diatoma 72
Dictyosphaerium 28
Dinobryon 66
Diplostauron 6
Draparnaldia 40

Enteromorpha 46
Euastrum 59
Eudorina 10
Euglena 89

Fragilaria 75

Gloeocapsa 97
Gloeotrichia 107
Gomphonema 82
Gomphosphaeria 100
Gonium 12
Gyrosigma 80

Haematococcus 5
Hyalotheca 51
Hydrodictyon 29

Lamprocystis 101
Lemanea 95
Lobomonas 6
Lyngbya 103

Mallomonas 65
Melosira 69
Meridion 73
Merismopedia 99
Mesotaenium 52
Micractinium 22
Micrasterias 57
Microcystis 101
Microspora 39
Mougeotia 34

Navicula 78
Nitzschia 83
Nostoc 105

Ochromonas 64
Oedogonium 35
Oocystis 16
Ophiocytium 60
Oscillatoria 103

Palmodictyon 47
Pandorina 11
Pascherina 9
Pediastrum 24
Peridinium 88
Phacotus 7
Phacus 90
Phormidium 103

Pinnularia 79
Pleuroccocus 30
Pleurotaenium 53
Porphyridium 92
Prasiola 43
Pteromonas 4
Pyramimonas 8
Pyrobotrys 9

Rhizoclonium 42
Rhoicosphenia 82
Rivularia 109

Scenedesmus 26
Scytonema 108
Spirogyra 37
Staurastrum 56
Stephanodiscus 70
Stephanosphaera 13
Stichococcus 32
Stigeoclonium 44
Stigonema 110
Surirella 84
Synechococcus 98
Synedra 74
Synura 67

Tabellaria 71
Tetraëdron 18
Tetraspora 47
Tetrastrum 27
Tolypothrix 108
Trachelomonas 91
Trebouxia 31
Trentepohlia 49
Tribonema 61

Ulothrix 38
Uroglena 68

Vaucheria 63
Volvox 14

Xanthidium 58

Zygnema 36
Zygogonium 36

HER MAJESTY'S STATIONERY OFFICE

*Government Bookshops*

49 High Holborn, London WC1V 6HB
13a Castle Street, Edinburgh EH2 3AR
41 The Hayes, Cardiff CF1 1JW
Brazennose Street, Manchester M60 8AS
Southey House, Wine Street, Bristol BS1 2BQ
258 Broad Street, Birmingham B1 2HE
80 Chichester Street, Belfast BT1 4JY

*Government Publications are also available through booksellers*